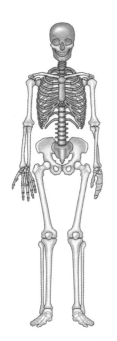

Human Anatomy
Flash Cards

Skeletal and Muscular Systems

Robert K. Clark
Cumberland County College
Vineland, New Jersey

1A
Skeleton, anterior view

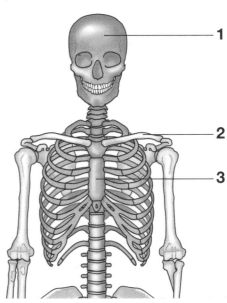

1
2
3

2A
Skeleton, anterior view

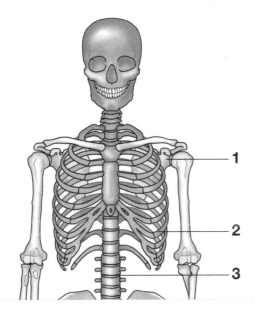

1
2
3

3A
Skeleton, anterior view

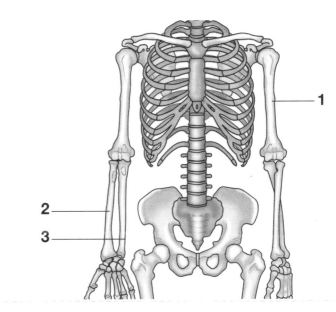

1
2
3

4A
Skeleton, anterior view

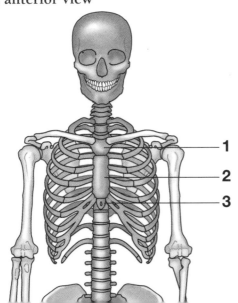

1
2
3

5A
Skeleton, anterior view

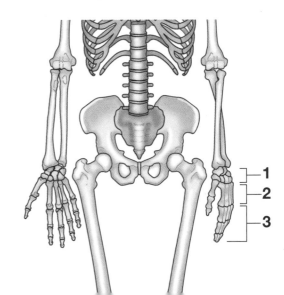

1
2
3

CONTENTS

1B

1. Skull
2. Clavicle
3. Sternum

3B

1. Humerus
2. Radius
3. Ulna

2B

1. Scapula
2. Ribs
3. Vertebral column

5B

1. Carpals
2. Metacarpals
3. Phalanges

4B

1. Manubrium
2. Gladiolus
3. Xiphoid process

6A
Skeleton, anterior view

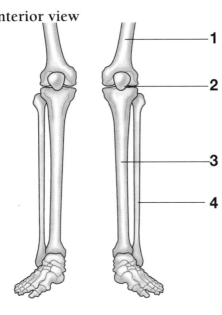

— 1
— 2
— 3
— 4

7A
Skeleton, anterior view

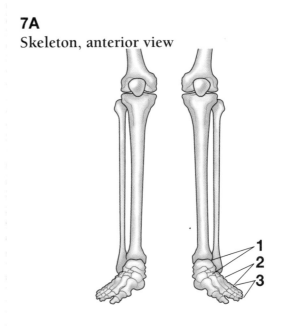

— 1
— 2
— 3

8A
Skull, anterior view

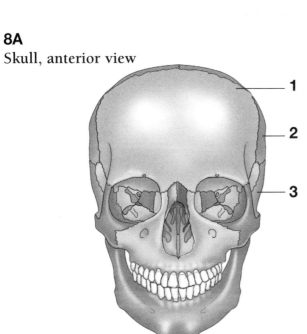

— 1
— 2
— 3

9A
Skull, anterior view

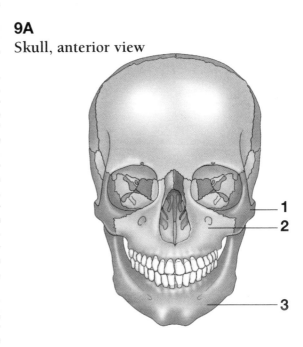

— 1
— 2
— 3

10A
Skull, anterior view

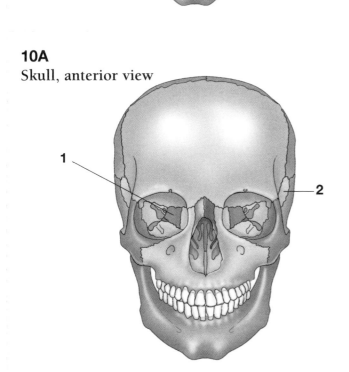

1
— 2

11A
Skull, anterior view

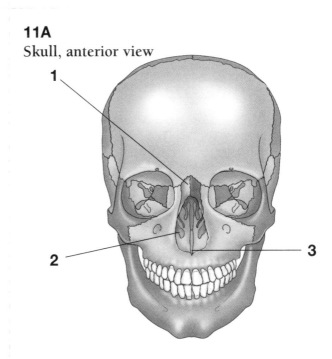

1
2
— 3

7B
1. Tarsals
2. Metatarsals
3. Phalanges

6B
1. Femur
2. Patella
3. Tibia
4. Fibula

9B
1. Zygomatic
2. Maxillary
3. Mandible

8B
1. Frontal
2. Parietal
3. Temporal

11B
1. Nasal
2. Nasal conchae
3. Vomer

10B
1. Ethmoid
2. Sphenoid

12A
Skull, lateral view

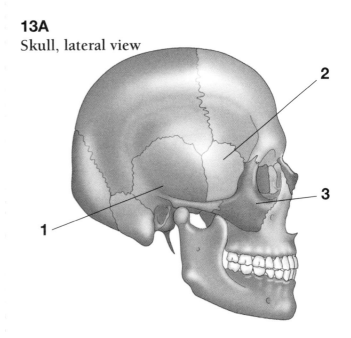

2 —

3 —

— 1

13A
Skull, lateral view

2

1

— 3

14A
Skull, lateral view

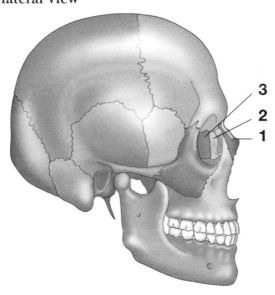

3
2
1

15A
Skull, lateral view

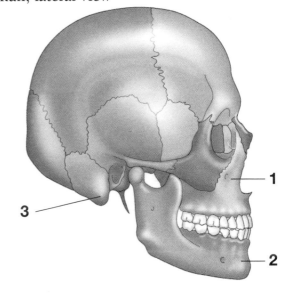

— 1

3

2

16A
Skull, lateral view

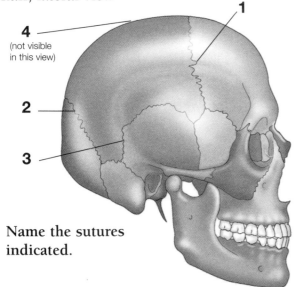

4
(not visible
in this view)

1

2

3

Name the sutures
indicated.

17A
Which skull bones contain paranasal sinuses?

13B
1. Temporal
2. Sphenoid
3. Zygomatic

12B
1. Frontal
2. Parietal
3. Occipital

15B
1. Maxillary
2. Mandible
3. Mastoid process

14B
1. Nasal
2. Lacrimal
3. Ethmoid

17B
1. Frontal
2. Sphenoid
3. Maxillary
4. Ethmoid

16B
1. Coronal suture
2. Lambdoid suture
3. Squamosal suture
4. Sagittal suture (Note: location shown, but suture not actually visible)

18A

1. The sella turcica is a feature of which bone?
2. What is located within the sella turcica?

19A
Vertebra, superior view

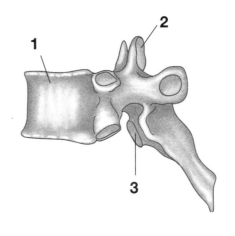

20A
Vertebra, superior view

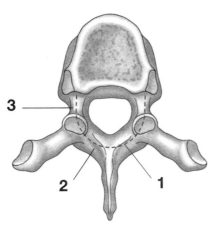

21A
Vertebra, lateral view

22A
Vertebra, lateral view

23A
Coxal bone, anterior view

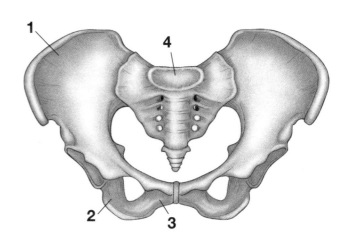

19B
1. Body
2. Vertebral Foramen
3. Spinous process
4. Transverse process

18B
1. The sphenoid bone
2. The pituitary

21B
1. Body
2. Superior articular facet
3. Inferior articular facet

20B
1. Arch
2. Lamina
3. Pedicle

23B
1. Ilium
2. Ischium
3. Pubis
4. Sacrum

22B
1. Intervertebral foramen
2. Spinous process
3. Transverse process

24A

1. What is the name of the socket that the head of the femur articulates with?
2. Which bones contribute to that socket?

25A
Femur, anterior view

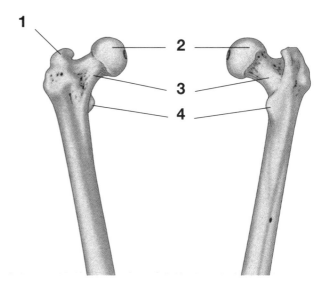

26A
Hand and wrist, anterior view

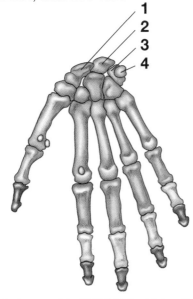

27A
Hand and wrist, anterior view

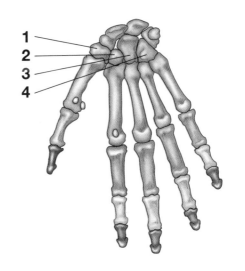

28A
Hand and wrist, anterior view

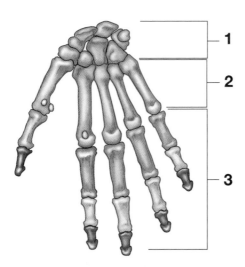

29A
Ankle and foot, superior view

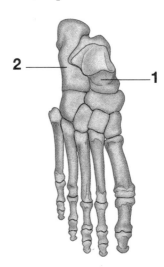

25B
1. Greater Trochanter
2. Head
3. Neck
4. Lesser Trochanter

24B
1. Acetabulum
2. Ilium, ischium, pubis.

27B
1. Trapezium
2. Trapezoid
3. Capitate
4. Hamate

26B
1. Scaphoid
2. Lunate
3. Triquetrum
4. Pisiform

29B
1. Talus
2. Calcaneus

28B
1. Carpals
2. Metacarpals
3. Phalanges

30A

Ankle and foot, superior view

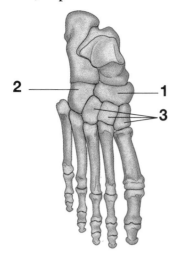

31A

Ankle and foot, superior view

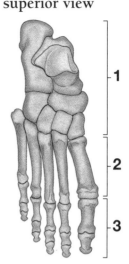

1A

Levator palpebrae superioris

Superficial muscles of the face and anterior trunk

2A

Masseter

Superficial muscles of the face and anterior trunk

3A

Orbicularis oculi

Superficial muscles of the face and anterior trunk

4A

Orbicularis oris

Superficial muscles of the face and anterior trunk

31B
1. Tarsals
2. Metatarsals
3. Phalanges

30B
1. Navicular
2. Cuboid
3. Cuneiforms

2B

Origin	Insertion	Action
Lateral zygomatic	Lateral mandible	Elevates mandible

1B

Origin	Insertion	Action
Sphenoid superior to eye	Upper eyelid	Elevates upper eyelid (opens eye)

4B

Origin	Insertion	Action
Connective tissue around mouth	Lateral margins of mouth	Closes lips, moves lips anteriorly

3B

Origin	Insertion	Action
Medial wall of orbit	Skin around eye	Closes eye

5A

Platysma

6A

Risorius

7A

External abdominal oblique

8A

External intercostals

9A

Internal intercostals

10A

Pectoralis minor

6B

Origin	Insertion	Action
Connective tissue of lateral face	Lateral margins of mouth	Moves margin of mouth laterally

5B

Origin	Insertion	Action
Connective tissue of shoulder	Mandible and connective tissue of face	Depresses mandible and lower lip moves lower lip posteriorly

8B

Origin	Insertion	Action
Inferior surface of each rib	Superior surface of each rib	Moves ribs superiorly and anteriorly

7B

Origin	Insertion	Action
Inferior 8 ribs	Connective tissue of anterior midline and lateral posterior ilium	Depresses ribs, compresses abdomen, flexes spine

10B

Origin	Insertion	Action
Ribs 3-5	Scapula	Protract and move shoulder inferiorly, rotate scapula inferiorly, elevate ribs

9B

Origin	Insertion	Action
Superior surface of each rib	Inferior surface of each rib	Moves ribs inferiorly and posteriorly

11A

Scalenes (group of 3)

12A

Rectus abdominus

13A

Serratus anterior

14A

Sternocleidomastoid

15A

Zygomaticus major

16A

Zygomaticus minor

12B

Origin	Insertion	Action
Anterior, superior pubis	Inferior costal cartilage and xiphoid	Depresses ribs, flexes vertebral column

11B

Origin	Insertion	Action
Cervical vertebrae	Superior surfaces of ribs 1and2	Elevate ribs, flex neck rotate head

14B

Origin	Insertion	Action
Medial clavicle and manubrium	Temporal	Flex neck, laterally flex and rotate head

13B

Origin	Insertion	Action
Anterior, superior surfaces of ribs 1-9	Anterior medial scapula	Protract shoulder

16B

Origin	Insertion	Action
Zygomatic	Superior lip	Elevates and retracts superior lip

15B

Origin	Insertion	Action
Zygomatic	Lateral margins of mouth	Elevates and retracts lateral margins of mouth

17A

Biceps brachii

Superficial muscles of the anterior shoulder and arm

18A

Brachialis

Superficial muscles of the anterior shoulder and arm

19A

Brachioradialis

Superficial muscles of the anterior shoulder and arm

20A

Deltoid

Superficial muscles of the anterior shoulder and arm

21A

Flexor carpi radialis

Superficial muscles of the anterior shoulder and arm

22A

Flexor carpi ulnaris

Superficial muscles of the anterior shoulder and arm

18B

Origin	Insertion	Action
Anterior distal humerus	Proximal ulna	Flex elbow

17B

Origin	Insertion	Action
Two sites on scapula	Proximal anterior radius	Flex elbow

20B

Origin	Insertion	Action
Clavicle and scapula	Lateral humerus	Abduct shoulder

19B

Origin	Insertion	Action
Distal lateral humerus	Distal lateral radius	Flex elbow

22B

Origin	Insertion	Action
Distal medial humerus and medial proximal ulna	Pisiform, hamate and metacarpal #5	Flex and adduct wrist

21B

Origin	Insertion	Action
Distal medial humerus	Metacarpals #2 and #3	Flex and abduct wrist

23A

Hypothenars (group of 3)

Superficial muscles of the anterior shoulder and arm

24A

Palmaris longus

Superficial muscles of the anterior shoulder and arm

25A

Pectoralis major

Superficial muscles of the anterior shoulder and arm

26A

Thenars (group of 4)

Superficial muscles of the anterior shoulder and arm

27A

Anconeus

Superficial muscles of the posterior trunk, shoulder, and arm

28A

Extensor carpi radialis brevis

Superficial muscles of the posterior trunk, shoulder, and arm

24B

Origin	Insertion	Action
Medial distal humerus	Connective tissue of palm	Flex wrist

23B

Origin	Insertion	Action
Pisiform, hamate and carpal tendons	Medial proximal phalanx of little finger and metacarpal #5	Abduct, flex and oppose little finger

26B

Origin	Insertion	Action
Scaphoid, trapezium, trapezoid, carpal tendons	Lateral metacarpal #1 proximal phalanx of thumb	Flex, adduct and oppose thumb

25B

Origin	Insertion	Action
Gladiolus, inferior medial surfaces of ribs 2-6	Proximal humerus	Flex, adduct and rotate shoulder medially

28B

Origin	Insertion	Action
Distal lateral humerus	Metacarpal #3	Extend and abduct wrist

27B

Origin	Insertion	Action
Distal lateral humerus	Proximal posterior ulna	Extend elbow

29A
Extensor carpi radialis longus

Superficial muscles of the posterior trunk, shoulder, and arm

30A
Extensor carpi ulnaris

Superficial muscles of the posterior trunk, shoulder, and arm

31A
Extensor digitorum

Superficial muscles of the posterior trunk, shoulder, and arm

32A
Infraspinatus

Superficial muscles of the posterior trunk, shoulder, and arm

33A
Latissimus dorsi

Superficial muscles of the posterior trunk, shoulder, and arm

34A
Levator scapuli

Superficial muscles of the posterior trunk, shoulder, and arm

30B

Origin	Insertion	Action
Distal lateral humerus and posterior ulna	Metacarpal #5	Extend and adduct wrist

29B

Origin	Insertion	Action
Distal lateral humerus	Metacarpal #2	Extend and abduct wrist

32B

Origin	Insertion	Action
Posterior scapula	Proximal posterior humerus	Rotate shoulder laterally

31B

Origin	Insertion	Action
Distal lateral humerus	Distal and middle phalanges of fingers	Extend fingers and wrist

34A

Origin	Insertion	Action
Cervical vertebrae 1-4	Medial scapula	Elevate scapula

33B

Origin	Insertion	Action
Inferior thoracic vertebrae, all lumbar vertebrae and all ribs	Proximal anterior humerus	Extend, adduct and rotate shoulder medially

35A

Rhomboideus (group of 2)

Superficial muscles of the posterior trunk, shoulder, and arm

36A

Splenius (group of 2)

Superficial muscles of the posterior trunk, shoulder, and arm

37A

Supraspinatus

Superficial muscles of the posterior trunk, shoulder, and arm

38A

Teres major

Superficial muscles of the posterior trunk, shoulder, and arm

39A

Teres minor

Superficial muscles of the posterior trunk, shoulder, and arm

40A

Trapezius

Superficial muscles of the posterior trunk, shoulder, and arm

36B

Origin	Insertion	Action
Vertebrae cervical 7– thoracic 6	Occipital, temporal. Vertebrae cervical 1-4	Extend, flex and rotate head

35B

Origin	Insertion	Action
Vertebrae cervical 7 and thoracic 1-5	Medial scapula	Adduct and rotate scapula inferiorly

38B

Origin	Insertion	Action
Inferior scapula	Proximal medial humerus	Extend, adduct and rotate shoulder medially

37B

Origin	Insertion	Action
Posterior scapula	Proximal posterior humerus	Abduct shoulder

40B

Origin	Insertion	Action
Occipital and vertebrae cervical 7 thoracic 1-12	Clavicle and scapula	Multiple movements of clavicle, scapula and neck

39B

Origin	Insertion	Action
Lateral scapula	Proximal lateral humerus	Rotate shoulder laterally

41A

Triceps brachii

Superficial muscles of the posterior trunk, shoulder, and arm

42A

Adductor longus

Superficial muscles of the anterior hip and leg

43A

Adductor magnus

Superficial muscles of the anterior hip and leg

44A

Gracilis

Superficial muscles of the anterior hip and leg

45A

Iliopsoas (group of 2)

Superficial muscles of the anterior hip and leg

46A

Pectineus

Superficial muscles of the anterior hip and leg

42B

Origin	Insertion	Action
Medial pubis	Posterior femur	Adduct, flex and rotate hip medially

41B

Origin	Insertion	Action
Scapula and humerus	Proximal posterior ulna	Extend elbow

44B

Origin	Insertion	Action
Inferior pubis	Proximal medial tibia	Adduct, flex and rotate hip medially

43B

Origin	Insertion	Action
Pubis and ischium	Posterior femur	Adduct, flex and rotate hip medially

46B

Origin	Insertion	Action
Superior pubis	Proximal medial femur	Flex, adduct and rotate hip medially

45B

Origin	Insertion	Action
Lateral vertebrae and ilium	Proximal medial femur	Flex trunk, flex and rotate hip laterally

47A

Peroneus longus

Superficial muscles of the anterior hip and leg

48A

Rectus femoris

Superficial muscles of the anterior hip and leg

49A

Sartorius

Superficial muscles of the anterior hip and leg

50A

Tensor fasciae latae

Superficial muscles of the anterior hip and leg

51A

Tibialis anterior

Superficial muscles of the anterior hip and leg

52A

Vastus lateralis

Superficial muscles of the anterior hip and leg

48B

Origin	Insertion	Action
Lateral anterior ilium	Proximal anterior tibia	Flex hip and extend knee

47B

Origin	Insertion	Action
Proximal lateral tibia and proximal fibula	Metatarsal #1 and #1 cuneiform	Plantar flex ankle, evert foot, support arch

50B

Origin	Insertion	Action
Lateral ilium	Lateral tibia	Flex and abduct hip

49B

Origin	Insertion	Action
Lateral anterior ilium	Medial anterior tibia	Flex knee, flex and rotate hip laterally

52B

Origin	Insertion	Action
Lateral proximal femur	Anterior proximal tibia	Extend knee

51B

Origin	Insertion	Action
Lateral proximal tibia	Metatarsal #1 and #1 cuneiform	Dorsiflex ankle invert foot

53A

Vastus medialis

Superficial muscles of the anterior hip and leg

54A

Biceps femoris

Superficial muscles of the posterior hip and leg

55A

Gastrocnemius

Superficial muscles of the posterior hip and leg

56A

Gluteus maximus

Superficial muscles of the posterior hip and leg

57A

Gluteus medius

Superficial muscles of the posterior hip and leg

58A

Peroneus brevis

Superficial muscles of the posterior hip and leg

54B

Origin	Insertion	Action
Posterior lateral ischium and posterior femur	Proximal fibula and tibia	Flex knee, extend and rotate hip laterally

53B

Origin	Insertion	Action
Posterior femur	Anterior proximal tibia	Extend knee

56B

Origin	Insertion	Action
Posterior lateral coxal, sacrum, coccyx	Posterior femur	Extend and rotate hip laterally

55B

Origin	Insertion	Action
Distal femur	Posterior calcaneus	Plantar flex ankle, invert and adduct foot, flex knee

58B

Origin	Insertion	Action
Fibula	Metatarsal #5	Plantar flex foot and ankle, evert foot

57B

Origin	Insertion	Action
Posterior lateral coxal	Proximal femur	Abduct and rotate hip medially

59A

Plantaris

Superficial muscles of the posterior hip and leg

60A

Popliteus

Superficial muscles of the posterior hip and leg

61A

Semimembranosus

Superficial muscles of the posterior hip and leg

62A

Semitendinosus

Superficial muscles of the posterior hip and leg

63A

Soleus

Superficial muscles of the posterior hip and leg

60B

Origin	Insertion	Action
Distal lateral femur	Proximal posterior tibia	Rotate tibia medially

59B

Origin	Insertion	Action
Distal femur	Posterior calcaneus	Plantar flex ankle, flex knee

62B

Origin	Insertion	Action
Posterior ishium	Proximal medial tibia	Flex knee, flex and rotate hip medially

61B

Origin	Insertion	Action
Posterior ishium	Proximal medial tibia	Flex knee, flex and rotate hip medially

63B

Origin	Insertion	Action
Proximal posterior fibula and tibia	Posterior calcaneus	Plantar flex ankle